15
14
1
2
3
4
7
4
5
6
REGIONS OF U. S.
AND CANADA

REGIONAL GEOGRAPHY OF THE UNITED STATES AND CANADA